Fatima Zahrae MRHARI
Mohammed RADOUANI

3D digitisation of heritage & robotic engineering

Fatima Zahrae MRHARI
Mohammed RADOUANI

3D digitisation of heritage & robotic engineering

Synergies between cultural heritage and advanced engineering

ScienciaScripts

Cover image: www.ingimage.com

This book is a translation from the original published under ISBN 978-3-8381-8960-4.

Publisher:
Sciencia Scripts
is a trademark of
Dodo Books Indian Ocean Ltd. and OmniScriptum S.R.L publishing group

120 High Road, East Finchley, London, N2 9ED, United Kingdom
Str. Armeneasca 28/1, office 1, Chisinau MD-2012, Republic of Moldova, Europe
Managing Directors: Ieva Konstantinova, Victoria Ursu
info@omniscriptum.com

Printed at: see last page
ISBN: 978-620-8-41270-8

Contents

3D DIGITISATION OF HERITAGE
& ROBOTICS ENGINEERING

Foreword

Technological advances in recent decades have transformed many sectors, opening up new perspectives in fields as diverse as cultural heritage and mechanical engineering. This project is part of this dynamic of innovation, exploring two distinct but complementary aspects: 3D digitisation applied to heritage and the design of a 6-axis robot using 3D printing.

3D digitisation, particularly in the field of heritage conservation, is opening up new approaches to the preservation and enhancement of historic monuments. It makes it possible to capture architectural and artistic details with great precision, guaranteeing faithful documentation of the works, while facilitating their restoration and study. In this project, the example of the Golden Mansour Gate in Meknes was used to illustrate the potential of these technologies in protecting and enhancing our cultural heritage.

At the same time, the development and manufacture of a 6-axis robot 3D printing technology (FDM) demonstrates the efficiency and flexibility of additive manufacturing methods. This process makes it possible to design complex mechanical systems while reducing costs and production times. Using an innovative approach, this project highlights the possibilities offered by 3D printing in the design of advanced robotic solutions.

This book offers an in-depth analysis of these two aspects, highlighting the key stages of digitisation and additive manufacturing, while exploring the technical challenges and solutions implemented to achieve the objectives set.

Summary

The project we carried out was divided into two distinct phases, each with its own technical and methodological objectives.

The first phase focused on a reverse engineering process applied to the monumental gates of Meknes, a cultural heritage requiring faithful reproduction. We began by scanning the doors to capture their complex geometry. The data obtained was then processed using specialised Geomagic Design and Wrap software, which enables surfaces to be reconstructed and modelled. Once the surfaces had been processed and optimised, we used a slicer to prepare the models for 3D printing. The use of this technology enabled us to reproduce the architectural details of the doors with great precision, while at the same time making the most of digital manufacturing methods.

The second phase of the project was devoted to the manufacture of a six-axis robot using Fused Deposition Modeling (FDM) technology. The aim of this part was to combine mechanical design and 3D printing to produce a robot capable of moving in six degrees of freedom, simulating complex robotic applications. The process involved several stages, including modelling the robot parts, simulating movements and validating mechanical feasibility. Using FDM printing, we were able to manufacture the various parts and assemble them, demonstrating the ability of this technology to produce functional and robust components for robotic applications.

These two phases of the project were particularly rewarding, as they enabled us to explore both artistic and technical fields, while using cutting-edge technologies. We not only strengthened our skills in reverse engineering, 3D printing and robotics, but also developed a better understanding of digital manufacturing processes applied to contexts as varied as cultural heritage and technological innovation.

General introduction

This project has two complementary strands, offering an opportunity to explore innovative technical and technological perspectives.

The first part is dedicated to the application of digitisation and reverse engineering technologies in the field of heritage. These tools open up new perspectives for the preservation, restoration and enhancement of cultural heritage. We have implemented 3D capture and digital modelling techniques to create faithful representations of emblematic heritage objects.

The second part focuses on the design and manufacture of a 6-axis robot prototype. To do this, we opted for additive manufacturing, more specifically 3D printing, to produce the mechanical components of this robotic system. This innovative approach enabled us to redefine the design of the robotic arm and explore new technical possibilities. In addition, a control programme based on the Arduino platform was developed to precisely control the coordinated movements of the various axes.

Although distinct, these two aspects are united by a common approach to technological innovation and exploration of the possibilities offered by digital tools. This project has enabled us to develop a global vision of the technical, scientific and societal issues associated with the use of modern technologies.

This report presents the main stages in our approach, the technical choices made and the results obtained for each part of the project.

Part 1

Digitisation and 3D printing in the heritage sector

Preserving our cultural heritage is one of the major challenges of our time. Faced with the deterioration of works and the difficulty of access for the general public, digitisation and 3D printing technologies represent tremendous opportunities. In this second part, we will explore how these technological innovations make it possible to faithfully capture the essence of heritage pieces, preserve them over time and share them more widely.

The preservation and enhancement of cultural heritage are major issues in contemporary society. Faced with the challenges posed by the degradation and destruction of many historic sites and artefacts, new emerging technologies offer innovative solutions for digitising and safeguarding this precious heritage, and the use of reverse engineering and 3D printing using stereolithography (SLA) is proving particularly relevant in this context.

This approach has many advantages for heritage preservation. It facilitates the conservation, study and display of objects, while reducing the risk of damage. What's more, SLA 3D printing offers high resolution, enabling the fine details of objects to be reproduced.

As well as preserving our heritage, digitising it also paves the way for greater accessibility and dissemination of knowledge.

In the following sections, we will detail the different stages of reverse engineering and SLA 3D printing applied to heritage digitisation, as well as the technical challenges and prospects offered by these technologies.

I. Industrial context

In the context of changing design constraints (shorter design times, increasing number of constraints, increasing number of function integrations, miniaturisation, etc.) and the development of new and innovative products, a company wishing to retain and/or acquire new market share must improve its control of the triptych: cost, lead time and quality.

- **Costs:** 80% of costs (for development and industrialisation) are incurred in the early stages of design (10 to 15% of total time), so design choices need to be validated as early as possible;
- **Quality:** the product must meet the requirements of sign, use and exchange, as well as producibility; any design errors leading to costly modifications and delays must be detected as early as possible;
- **Deadlines**: faced with market fluctuations, a company that wants to be competitive must be able to develop its new products more quickly. A delay of six months in bringing a product to market can reduce profits by 33%, whereas an overrun in the research budget would only result in a loss of 5%.

To meet these essential success criteria, companies have had to adapt their design processes, but also take into account the emergence of new technologies.

Modern systems have reached such a level of sophistication that it would have been difficult to imagine using traditional methods. In addition, the emergence of new working methods such as concurrent engineering means that all stages of a product's life cycle can be taken into account, and the product and its means of production can be developed jointly. Four main concepts characterise concurrent engineering:

- **Simultaneity**: the parallel involvement of activities (and tasks), services (and professions) reduces waste and wasted time, in particular by avoiding excessively long lead times;
- **Competition:** by exploring different variants of a product in order to select the solution best suited to the problem posed;
- **Breakdown of the project into sub-projects with clearly identified interfaces**;
- **Integration:** designing a product requires a wide range of professional skills (marketing people, ergonomists, electronics engineers, mechanics, methods, etc.), brought together in a project group that is present throughout the product life cycle.

II. Deep ExplorationReverse Engineering at the Analysis and Innovation Department

Reverse engineering is a systematic process by which an object, system or product is analysed in order to understand its inner workings, structure and the methods used in its design or manufacture. This process generally involves collecting data on the object under study, such as measurements, images or technical specifications. This data is then analysed and interpreted to reconstruct or model the object in whole or in part. To explore this reverse engineering process in more detail, let's look at the main stages involved:

1. 3D scanning

(1) Introduction

Recent advances in computer technology have significantly changed the way we perceive objects. In various industrial processes, a simple photo is no longer sufficient, and the need to recover the geometry of an existing real product in 3D is increasingly frequent. 3D digitisation is a technique that enables the creation and description of a digital model of a "physical object" in the form of a "point cloud".

The design/industrialisation/manufacture of mechanical products is no longer the only field in which these principles are used. Nowadays, measuring objects in 2D, and even more so in 3D, has become an indispensable tool in many related fields: automotive, aeronautics, electronics, medical, heritage, luxury goods, etc.

The market for this technique is still developing and is set to expand significantly over the next few years in a wide range of applications: *reverse engineering, shape copying, aesthetics, biotechnology, inspection, etc.*

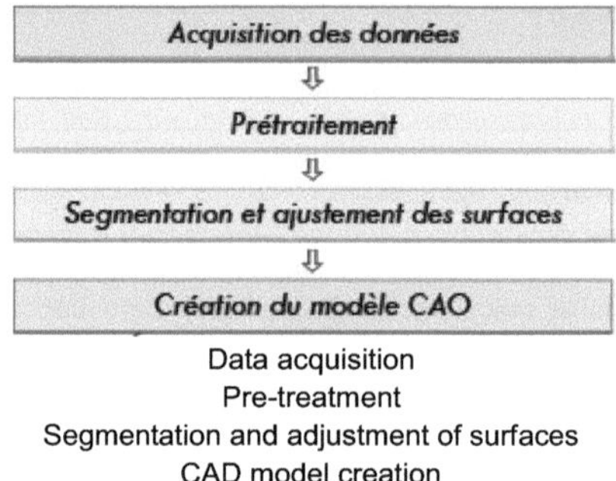

Data acquisition
Pre-treatment
Segmentation and adjustment of surfaces
CAD model creation

Figure 1: Basic steps in reverse engineering

Today's 3D scanning systems can quickly and accurately capture data at very high densities, enabling even the smallest surface details to be represented. However, the data collected is far from ideal, and suffers from various problems such as a lack of information about certain regions, noise from different sources, multiple views, etc. Given the complexity of the shapes digitised, the acquisition used and the intended application, the digital image acquired requires a certain amount of pre-processing to make it usable (Figure 1).

The aim of this section of Appendix 3 is to highlight the different characteristics of 3D scanning systems.

(2) Digitisation chain

The digitisation chain is made up of hardware and software components: an acquisition system, a movement system and a processing system. The quality of the data and information in terms of accuracy is dependent on the performance of of each of these elements, which must be consistent with each other in order to remain as faithful as possible to the initial shape of the object (Figure 2).

Figure 2: 3D scanning chain

a. Acquisition systems

Acquisition systems enable the scene to be observed and digitised. They use a wide range of technologies and are generally referred to as 3D sensors. Acquisition can be classified according to the type of sensors used: *contact or non-contact sensors.*

b. Contact 3D sensors

Contact sensors are derived from three-dimensional metrology and require a displacement system. They consist of a rigid stylus that detects mechanical contact with the surface of the object and records the Cartesian coordinates of the point being probed. There are two different contact probing techniques: discrete 'point-by-point' with a dynamic trigger sensor (Figure *3a*) and 'shape-following' with a continuously scanning sensor (Figure *3b*).

Contact probing offers the advantage of very high accuracy. The force exerted by the stylus on the object is low, "of the order of 0.02 N/mm^2", but this can affect the capture of points on highly deformable parts. The limitation of such a digitisation system lies in the speed acquisition

Figure 3: Renishaw point-to-point and shape-following probes tracking" probes

With the point-by-point sensor, it generally takes several days to scan a part with a few hundred thousand points. The shape-following sensor, on the other hand, allows dense scanning in reasonable times, but requires the use of an automatic movement system.

c. Non-contact 3D sensors

Non-contact 3D sensors are very diverse, and the technologies and physical principles used are varied (Figure 4) and (Figure 5). They can be grouped according to their properties: active or passive if the environment is illuminated or not; single- or multi-ocular if the system produces one or more images.

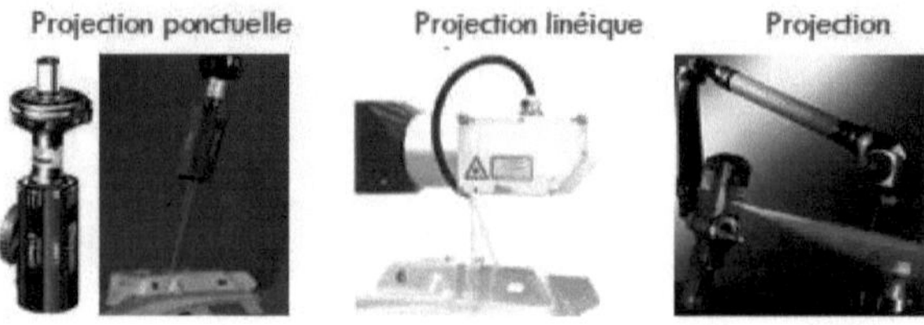

Figure 4: Scanning a surface using a non-contact scanning system

Figure 5: Portable 3D laser scanner for large products

The most commonly used techniques are triangulation, telemetry and image processing.

A Laser triangulation: consists of arranging a light source (usually a laser beam), the object to be digitised and a sensor (a Charge Coupled Device camera) in a triangle. Knowing the characteristics of the transmitter and receiver and interpreting the reflected radiation enables 3D information to be obtained.

A Telemetry consists of measuring the distance between the object and the transmitter-receiver in direction of transmission-reception of a wave. The wave can be of different types: laser "light wave"; sonar "ultra sound wave"; radar "radio wave".

A Image processing is a fully-fledged technique in the field of 3D digitisation. Its major advantage over triangulation or telemetry techniques is that it provides the internal structure of the object, which is absolutely essential in medical imaging (X-ray tomography, magnetic resonance imaging, ultrasound scanners, etc.).

(3) Travel systems

In practice, limitations due to the accessibility and field of view of 3D scanning equipment often mean that the number of degrees of freedom in translation and/or rotation of the sensor in relation to the object must be increased. Two main types of structure are used to reach all positions in space:

A "Cartesian" structures (Figure *6a*): typically a coordinate measuring machine plus additional rotations added to the sensor (motorised head or manual positioning block) and to the part by placing it on a turntable.

i "Polar" structures (Figure *6b)*: polyarticulated arms, robots, turntables,

(a) Machine à mesurer

(b) Bras-articulé

Figure 6: Example of a carrier performing the "move" function

2. 3D Modelling

The purpose of this section of Appendix 3 is to present the pre-processing phase of the digitised point clouds and to assess the performance and limitations of the filtering and pre-modelling functions when dealing with noisy models.

2.1 Treatment systems

The processing systems manage the synchronisation of the displacement system with the acquisition system and process the data from the sensor for appropriate formatting. The function of the processing software is establish the transition relationship between the corresponding 1D/2D and 3D data. It is also used to carry out the operations associated with pre-processing: resetting views, cleaning up data, qualifying points, etc.

It is at the level of the processing system that the user can introduce certain parameters reflecting his requirements relative to the specifications. He chooses the scanning pitch, the minimum resolution and also the type of data he wants to obtain. In this way, they establish their digitisation strategy.

2.2 - Pre-processing of digitised point clouds

Data from a non-contact 3D scanning system is often discrete, dense, inhomogeneous and noisy. The pre-processing stage includes operations such as :

- Resetting or consolidating views.
- Data filtering or cleansing.

a. Relocating several views

In order to obtain a complete image of the object's surface, it is necessary to scan several views. The re-alignment or consolidation of views consists of expressing all the scanned point clouds in the same frame of reference. If we consider a point cloud n1 from view 1 expressed in frame R1 and a point cloud n2 from view 2 expressed in frame R2, rescanning enables us to define the transformation to be applied to n1 so as to bring part of the data from the latter into coincidence with n2 in their overlap zone.

i. Manual resetting

Manual registration consists of asking the operator to digitise, in each view, specific geometric elements, such as registration spheres, at the same time as the object, or to designate the overlap zones on each point cloud, based on the object's remarkable characteristics. Transformation matrices are then calculated to place all the data in the same frame of reference (Figure 7).

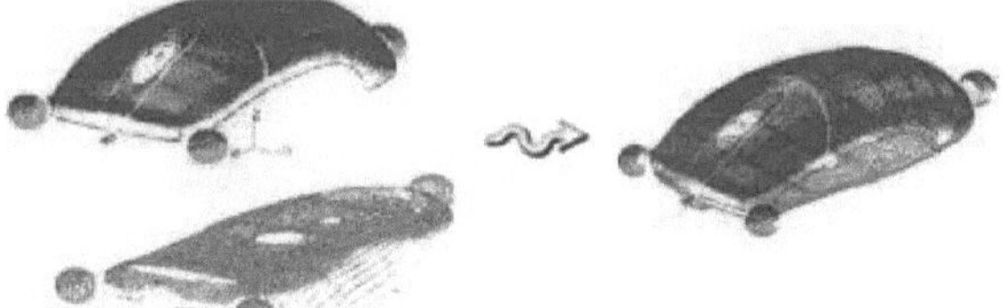

Figure 7: Mapping using ideal spheres

ii. *Automatic resetting*

Automatic registration involves automatically matching overlapping views by recognising and mapping common geometric features. The ICP method[1] is commonly used, as it combines both automatic matching (associating any point on the first feature with the closest point on the second feature) and transformation identification (least squares optimisation of the displacements that best align the matched features).

b. Filtering digitised data

Various causes of noise can affect the digitisation process: optical noise due to the imperfection of the lenses making up the camera, electronic noise due to the imperfection of the components and vibratory mechanical noise due to the different operating clearances in the links. In most cases they

[1] Iterative Closest Point

appear in the form of rough portions and scattered points on either side of the discretised surface, known as "out liners". They can be treated using several methods:

A Hardware filtering: integration of filters appropriate to the nature of the noise in the digitisation phase.

A Software filtering: development of algorithms for detecting and filtering noisy data in the point clouds generated.

A Hardware-software filtering: Total or partial combination of the first two methods.

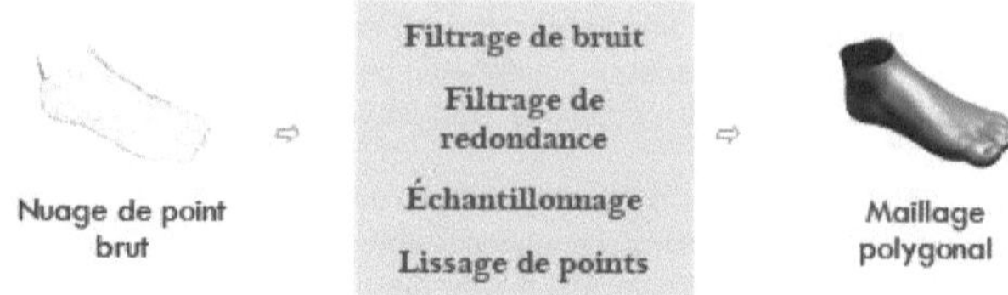

Raw point cloud

Noise filtering Redundancy filtering Sampling Point smoothing

Polygonal mesh

Figure 8: Pre-processing phase: noise filtering

The most widely used commercial software in this field is : Rapidform, Geomagic, the Spider module in ALIAS Studio, Polyworks and the Digitized Shape Editor module in CATIA. They offer functions for retrieving, viewing, editing and manipulating dense point clouds quickly and easily.

c. Meshing of filtered models

Meshing is a pre-modelling in which we move from an unstructured point cloud to a model that is well structured from a geometric and topological point of view.

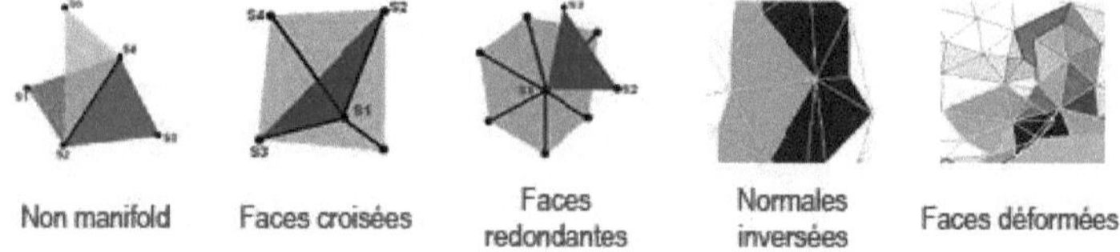

Non-manifold Crossed faces Redundant faces
Inverted normals Deformed faces

Figure 9 Types of abnormal faces.

Figure 9 illustrates the types of abnormal faces encountered on a mesh. Rapidform and Polyworks software offer functions for detecting these anomalies.

Comparing the meshes of the filtered models with the raw reference model reveals their degree of concordance and enables us to estimate whether the mesh can generate a correct segmentation. The illustration uses the Stanford rabbit available in the literature to highlight the quality of the meshers implemented in the most widely used commercial software: Rapidform, Geomagic, the Spider module in ALIAS Studio, Polyworks and the Digitized Shape Editor module in CATIA (Figure 10).

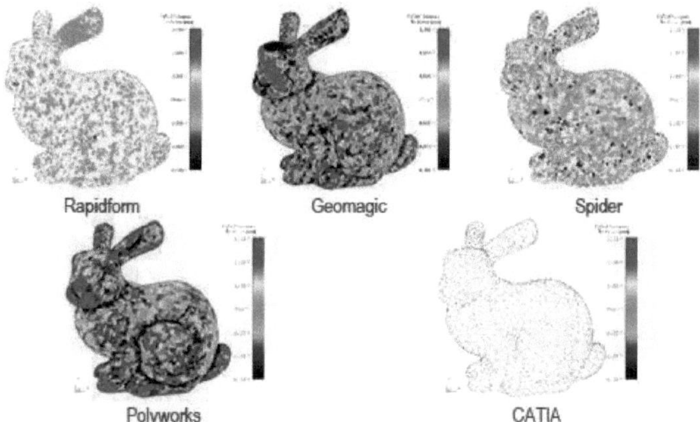

Figure 10: Comparison of polygonal models with the filtered point cloud

2.3 Partitioning or segmenting a point cloud

a. Structuring or pre-modelling data

The pre-modelling stage is used to structure the data. The possibilities : the polyhedral model, the surface model and the
of spatial occupation (Figure 11).

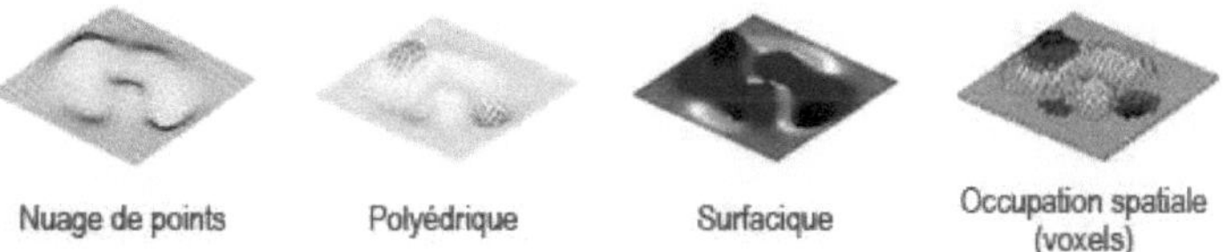

Polyhedral point cloud Surface Spatial occupancy (voxels)

Figure 11: Data representation models

Triangular meshes are the most widely used because they represent a preferred solution for industry due to their simplicity and flexibility.

b. Segmentation of a point cloud

Segmentation is the process of partitioning the mesh into homogeneous regions. It has emerged as an important issue due to the complexity of its implementation and the multitude of applications that depend on it. Segmentation methods can be classified into two categories:

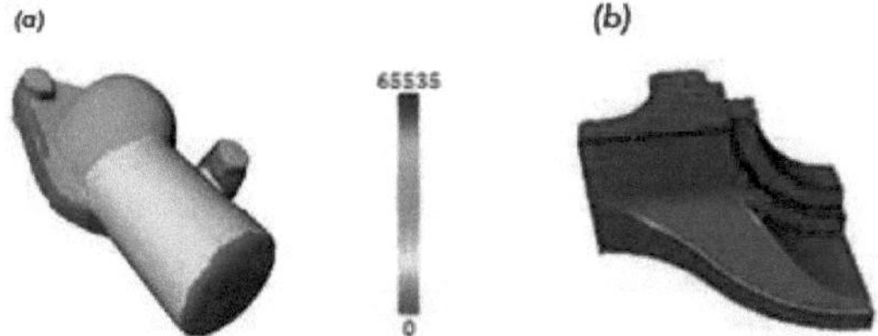

Figure 12: Region growth (a) and contest detection (b

A Region identification techniques can be used to partition the triangular mesh into homogeneous surface patches (Figure *12a*).

A Contour detection techniques look for discontinuity points in local features. In 2D image processing, contours high frequencies and detected by analysing the image gradient. For a 3D mesh, a contour can be detected by analysing the orientation of the normals (Figure *12b*).

Hybrid segmentation (or cooperative border/region segmentation) is used to partition the preprocessed point cloud into homogeneous subsets. It consists of locating the contours that delimit the different regions in the point cloud using a frontier approach. The information obtained on the position of the contours will be used reinforce the stopping criteria in the region growth approach.

3. Rapid prototyping

3.1-Definition

A new industrial revolution with so-called 3D printers or rapid prototyping machines (Figure 13).

Figure 13: Examples 3D printers

Rapid prototyping technologies open the door to rapid manufacturing processes that can be implemented directly and locally, making it possible to produce objects rapidly in small series and with many variants, which is not possible with more traditional technologies. In addition, 3D printers are capable of producing highly complex parts that are impossible to mould or machine, with several components already assembled and manufactured in a single step (Figure 14).

Figure 14 :Rapid prototyping: 3D printed models and objects

The term "rapid prototyping" covers a range of processes that physically reproduce a 3D object from a plan transcribed into a CAD file. The various rapid prototyping techniques mean models can be obtained in a very short space of time, enabling the design to be validated from an aesthetic, geometric, functional or technological point of view. Rapid prototyping integrates three essential concepts: time, cost and complexity of shapes.

- Time: the aim of rapid prototyping is to produce models quickly, in order to reduce product development times.
- CoOt: rapid prototyping makes it possible to produce prototypes without the need for expensive tooling, while guaranteeing the performance of the final product. This means that different variants of the product being developed can be explored in order to select the most appropriate solution.
- Complex shapes: machines that material are capable of producing extremely complex shapes (inclusions, cavities, etc.) that cannot be achieved by processes such as machining.

3.2-IT resources

Object

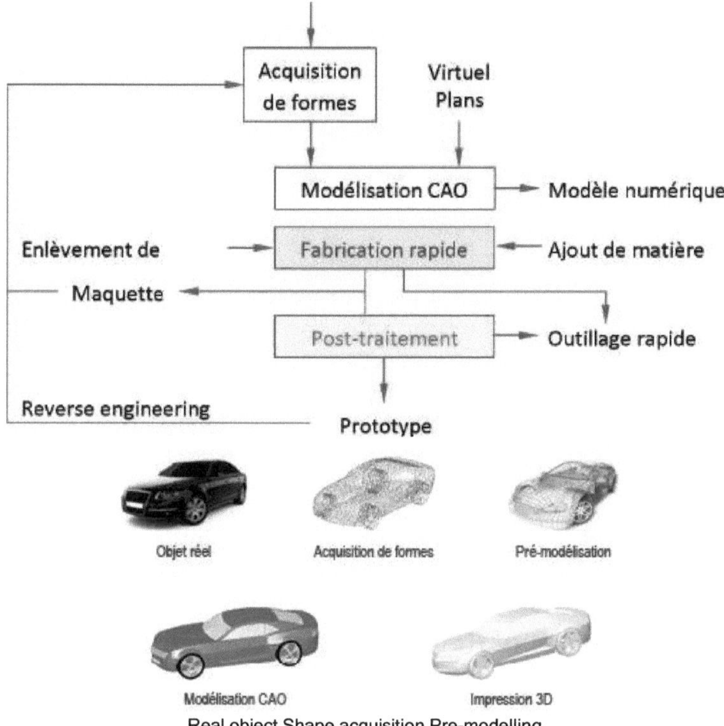

Real object Shape acquisition Pre-modelling
CAD modelling 3D printing

Figure 15: Rapid prototyping process

The IT resources that are becoming increasingly important in rapid prototyping are (Figure 15):

- reverse engineering (shape acquisition systems combined with surface reconstruction software);
- Computer-aided design (CAD), manufacturing processes involving the addition and removal of material;
- Post-processing, such as duplication using silicone moulds, investment casting, etc.

3.3-Materials

Raw material systems have been developed for use with printers
3D. Unused powder can be recycled to reduce printing costs.

Figure 16: Materials

3.4-Steps in producing a model

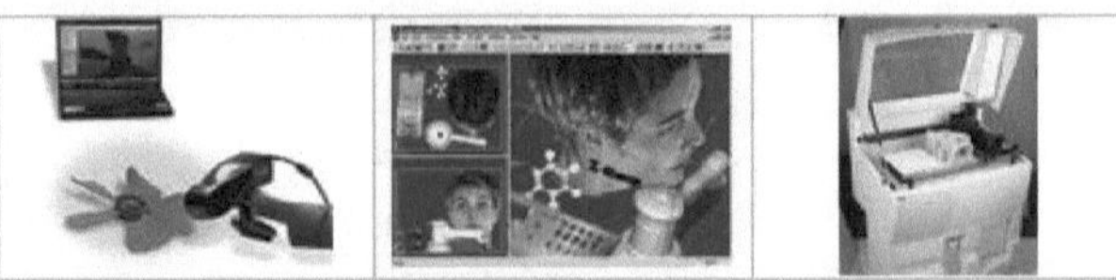

No toxic products Cellulose powder Plastic powder Gypsum powder Ceramic powder

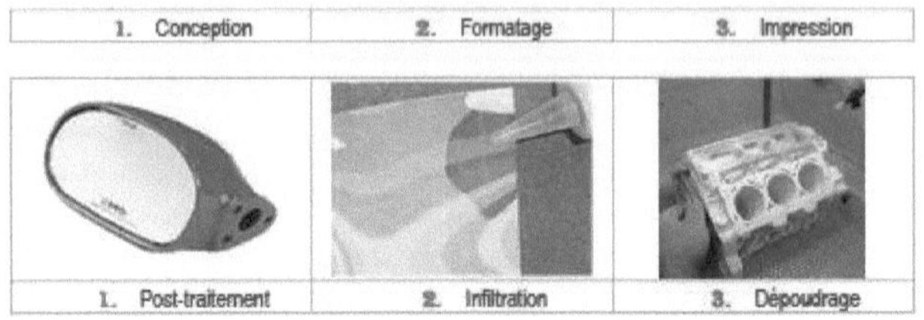

*Design **2.** Formatting **3.** Printing*
***1.** Post-treatment **2.** Infiltration **3.** Desoldering*
Figure 17: Materials

III. The Many Faces of Reverse Engineering: Innovative Applications in Diverse Fields

Although often associated with product design, reverse engineering also has practical applications in a variety of fields. In the manufacturing industry, for example, it is used to analyse and improve existing products by identifying design flaws and proposing innovative solutions. In mechanical engineering, reverse engineering is invaluable for reverse-engineering obsolete or unavailable parts, thereby extending the life of equipment and reducing replacement costs.

1- Reverse engineering and surface reconstruction

The reverse engineering of environments and the reconstruction of surfaces are two application domains that are similar in their methodology but nevertheless distinct in their purpose. The aim is to obtain CAD models that 'stick' as closely as possible to the point cloud.

1.1- Reverse engineering of environments

The reverse engineering of environments involves obtaining CAD models of large-scale, complex man-made environments, such as civil or industrial architecture (Figure 18). In this case, modelling is most often done with what we call "simple geometric primitives": plane, cylinder, torus, sphere, cone.

Figure 18: Dupont de Nemours chemical plant, Victoria (MENSI image)

1.2- Surface reconstruction

Surface reconstruction in geometric modelling is a process involves modelling mathematical surfaces (more complex surfaces: Splines, NURBS, Bézier surfaces) or reproducing manufactured objects identically.

Figure 19: Surface reconstruction and 3D printing (Boeing 777)

Surface reconstruction is used today in the design of many objects, both domestic (hoovers, telephones, toys, etc.) and industrial (cars, aeronautics, etc.). (Figure 19).

2- Computer-aided manufacturing CAM

Manufacturing is of course essential for the final production of a product. Recent computer-assisted manufacturing techniques use digitised surfaces of parts to copy shapes and adapt machining paths. (Figure 20).

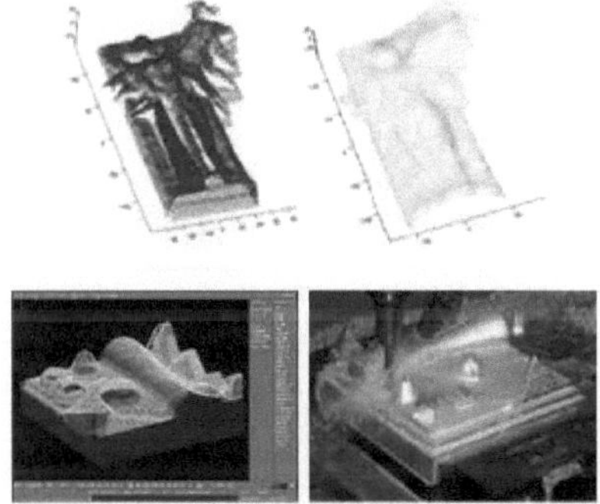

Figure 20: Example of machining paths for digitised CAD models

3- Biotechnology

Many 'orthotists use scanners to record a patient's shape. This technique is gradually replacing the time-consuming plaster cast. CAD/CAM software is used to create and manufacture orthoses or prostheses.

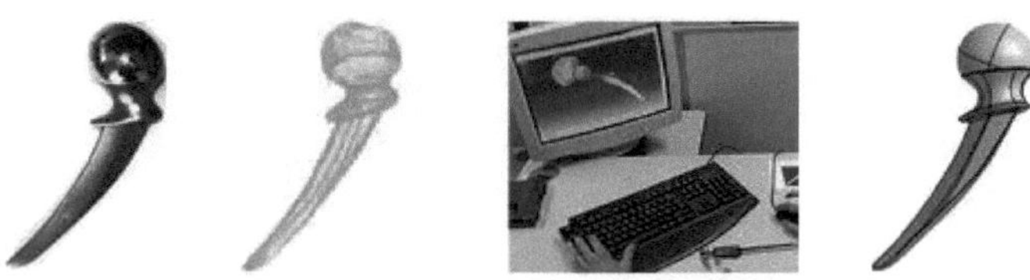

Figure 21: Application to the femoral head of a total hip prosthesis

4- Cultural heritage

One way of obtaining three-dimensional information about a building's heritage is to redesign the parameterised geometry of non-existent structures from manuscripts using CAD tools, and to model and dimension the existing heritage using digitisation and prototyping systems.

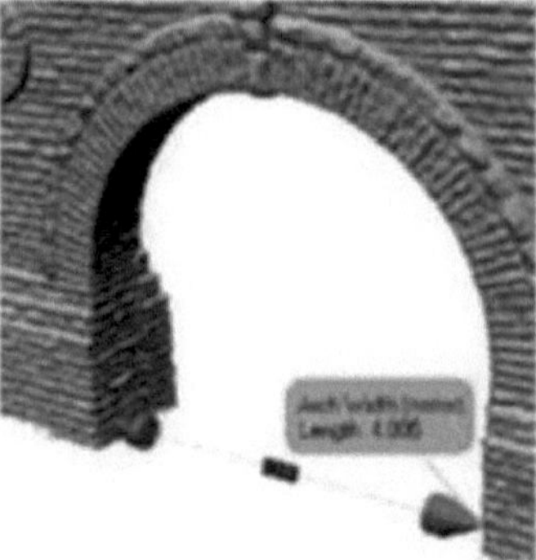

Figure 22: Point cloud model of a door

In this context, one of the aims of this project is to use 3D image analysis to determine the impact of structural restoration modifications on weathering phenomena. [1]

5- Entertainment

3D scanners are used by the entertainment industry to create 3D models for films and video games. If a physical representation of a model exists, it is much quicker to scan it than to create it manually using 3D modelling software. Often, artists will sculpt a physical model and scan it rather than making a digital representation of it directly on a computer.

In addition to the case studies presented, this project also provided opportunity to explore 3D scanning and additive printing techniques in a more generic context. The first step in discovering these methods was to laser scan a simple desk mouse. This reverse engineering approach generated a detailed point cloud of the object. This point cloud was then processed using Geomagic Design software to reconstruct a faithful digital 3D model in the form of an STL file ready for printing. This preliminary experiment not only familiarised the team with the entire workflow, data acquisition to manufacturing, but also demonstrated the versatility of these technologies for digitising objects of varying size and complexity. The exercise thus laid a solid foundation for confidently tackling the more ambitious cultural heritage preservation challenges presented in the case studies. [2]

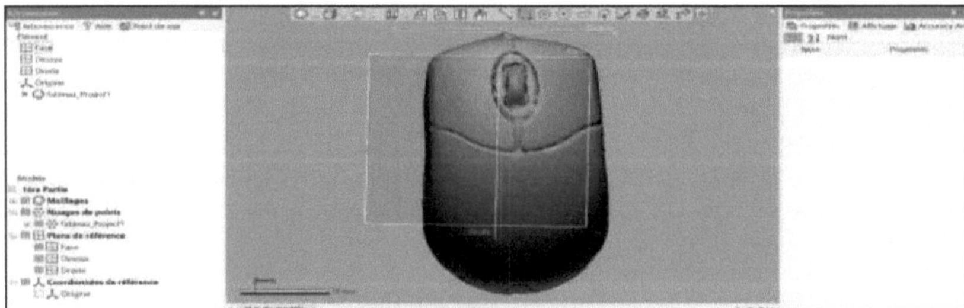

Figure 24: Analysis of the model by Geomagic Design

Figure 23: Mouse processing by Geomagic Design

IV. Virtual Conservation: The Transformative Role of Reverse Engineering in Cultural Heritage

In the field of heritage, reverse engineering is of particular importance for the preservation and enhancement of cultural and historical assets. Using advanced digitisation techniques such as photogrammetry and laser scanning, it is possible to create accurate and detailed digital representations of monuments, artefacts and historic sites.

These digital models make it possible to meticulously document the architectural, artistic and historical details of heritage works, facilitating their long-term conservation. They also offer unique opportunities for research, education and tourism promotion.

Historians and archaeologists can use these models to study the evolution of monuments, reconstruct their history and better understand their cultural significance.

Educators can integrate these models into interactive educational programmes, providing students with an immersive and engaging experience of heritage discovery. Finally, digital models can be used to create virtual tours, mobile applications and other digital tools, enabling visitors around the world to discover and appreciate cultural heritage in innovative and accessible ways. [3]

Real-life examples illustrate this transformative approach: the reconstruction of Notre-Dame Cathedral in Paris, which was severely damaged by fire in 2019, relies on reverse engineering to precisely restore its emblematic architectural elements. Similarly, the world heritage site in China has been rebuilt using 3D printing. In fact, 3D printing has been used to produce a reproduction of one of the Yungang caves. The site is a Buddhist temple with a multitude of caves included on UNESCO's list of world heritage monuments. It was excavated in the 5th-6th century AD and comprises a total of 252 caves, 51,000 statues and a sculpted area of 18,000 square metres.

Unesco makes no secret of their importance. According to them, the caves "constitute a classic masterpiece of the first peak of Chinese Buddhist art", allowing visitors to relive the history and spirituality of this ancient site.

In short, reverse engineering plays an essential role in preserving and promoting heritage by offering innovative ways of documenting, studying and sharing the cultural treasures of the past for future generations. [4]

Figure 25: 3D model of Notre Dame de Pris cathedral

Figure 26: The World Heritage site in China

V. Reverse Engineering and 3D Digitisation for Architectural Heritage

1. Reverse engineering applied to the door of the Mohammed 5 Museum in Rabat

In the application of reverse engineering to Morocco's historic monuments, the Mohammed 5 Museum Gate in Rabat is an emblematic example of the country's architectural and cultural wealth. Historically, this majestic gate bears witness to Moroccan architectural ingenuity and traditional craftsmanship. Its impressive architecture, characterised by carved stone arches and intricate decorative motifs, makes it a major historical symbol of the region.

As part of our reverse engineering project, the 3D digitisation of the Mohammed 5 Museum Gate is of particular importance. We plan apply data capture techniques, such as photogrammetry using an application called KIRI ENGINE, to document every architectural detail of this emblematic structure. The first steps will be to carry out precise surveys in the field, capturing images of the gate in its environment.

Figure 28: Photogrammetric scan of the door Figure 27: Application used for scanning

This data then processed and merged to create a detailed, true-to-life 3D model of the Mohammed 5 Museum Gate.

As well as capturing the shapes and dimensions, our project also included restoring the door's characteristic textures and decorative details. Using advanced post-processing techniques, we were able to recreate the complex patterns and traditional ornamentation that adorn the surface of the door, preserving its original authenticity and aesthetic thanks to Geomagic Design, Geomagic Wrap and Meshmixer software.

Figure 29: Geomagic Design point cloud processing

As a result, our digital model of the Mohammed 5 Museum Gate will not only provide an accurate visual representation of this historic structure, but will also serve as a valuable tool for its preservation, documentation and dissemination to the public, helping to promote Morocco's cultural heritage.

Following on from our reverse engineering and 3D scanning approach, once we had obtained the digital model in STL format, we began to process it using slicer software with a view to 3D printing. This slicing process is of crucial importance in preparing the model for additive manufacturing. The slicer is used to segment the model into successive layers, thus determining print paths and parameters required to produce each layer.

This step is essential to ensure the quality and accuracy of 3D printing, taking into account factors such as resolution, fill density, and any supports required. [5]

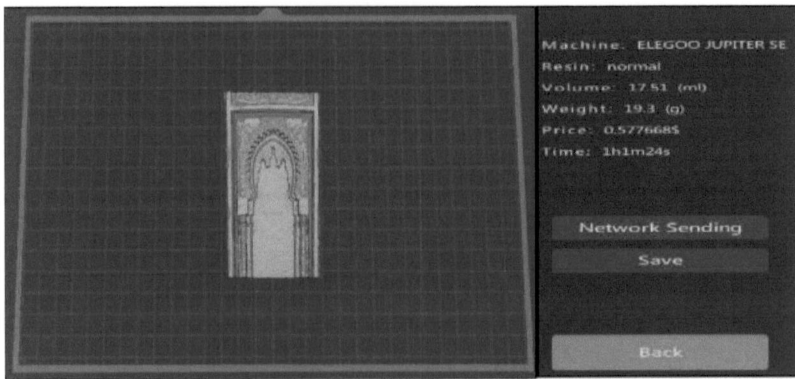

Figure 31 Preparing the model for 3D printing

Figure 30: Final result of the door after printing

The end result is a striking physical reproduction of this heritage piece, accurately capturing the essence and finesse of the original architectural details. This 3D-printed version thus offers an innovative solution for safeguarding and promoting Morocco's rich cultural heritage.

2. Preserving architectural heritage with 3D printing: from Morocco to Rome's Colosseum

Building on the experience gained from the successful 3D printing of the historic Moroccan Gateway, we decided to take on a new challenge by importing the 3D model of the majestic Colosseum in Rome. This iconic architectural monument, built in the 1st century AD under Emperor Vespasian, presented even greater levels of detail and complexity than the previous project. However, thanks to the gradual mastery of the 3D printer's key parameters, in particular the use of stereolithography (SLA) technology, which offers exceptional resolution and surface finish, we were able to produce a physical reproduction of breathtaking quality. The final result captures with remarkable fidelity the imposing structure of the Colosseum, with its arches, columns and delicate ornamentation. Almost 50 metres high and able to seat up to 80,000 spectators, this masterpiece of Roman architecture has survived the centuries to become one of the world's most recognisable symbols. This technological feat has enabled us to demonstrate the potential of reverse engineering and 3D printing to preserve and enhance the building's heritage in a way that makes it unique.
unprecedented level of detail the treasures of the world's architectural heritage.

Figure 33: Real image of the Colosseum in Rome Figure 32: Digital model of the Colosseum

After importing the 3D model of the Colosseum from the Thingiverse website, where it was available as a detailed point cloud, we used Chitubox software to carefully prepare the model for 3D printing. Using this preparation tool, we were able to fine-tune key parameters such as resolution, supports and model orientation, achieve the best possible result when we finally printed this iconic piece. [6]

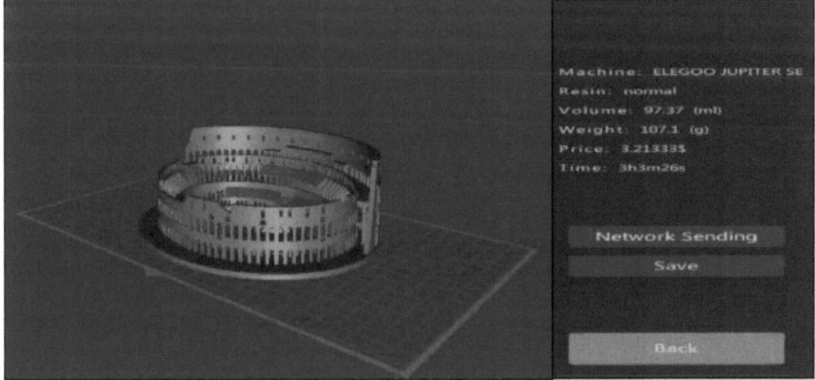

Figure 35: Preparing the model using the Chitubox slicer

Figure 34: Final result of the Coliseum after 3D printing

3. Challenges of the 3D reproduction of the Coliseum versus the door of the Mohammed V Museum

Unlike our previous project on the historic doorway of the Mohammed V Museum, where we used photogrammetry, a relatively simple process but limited in the capture of detail, the Colosseum in Rome presented a far greater challenge terms of reverse engineering. This iconic structure required highly accurate 3D data to faithfully reproduce its complex architecture. We therefore used a digital model derived from specialised 3D scans and advanced photogrammetry. Despite the quality of this data, the impressive size and rich ornamentation of the Colosseum presented a real challenge when preparing the digital model for 3D printing. The use of Chitubox software proved crucial in fine-tuning key parameters such as resolution, supports and model orientation, in order to optimise the quality of the final rendering and faithfully reproduce the finest architectural details. Managing these different aspects of reverse engineering in Chitubox required painstaking work, but enabled us to achieve a 3D printing result that was equal to the grandeur of the Colosseum.

V. Creation of a digital library of the monumental gates of Meknes

The successful experience of reverse engineering and 3D printing the door of the Mohammed V Museum in Rabat, as well as the Colosseum in Rome, has demonstrated the potential of reverse engineering and digitisation to preserve and enhance architectural heritage. Building on these encouraging results, the final part of this project focuses on the creation of a digital library of the monumental gates of Meknes.

This initiative is part of a strategy to safeguard and promote the cultural heritage of this historic city. The urban landscape of Meknes is dotted with numerous old gates of great architectural and historical value, but their varying state of conservation requires particular attention.

Digitising these heritage features will ensure that they are accurately documented and preserved in the form of 3D models, while promoting them to the public.

This final part of the project therefore aims to put in place a proven reverse-engineering methodology, in order to create an exhaustive digital library of the main monumental gates of Meknes.

This digitisation work will constitute a valuable heritage database, facilitating studies, restoration and promotion of this rich architectural heritage. [7]

1. Pilot project: Digitisation of the doorway of a traditional Riad

Before undertaking the creation of the digital library of the city's main doorways, an initial experiment was

carried out on the doorway of a Riad located in the old medina of Meknes. This door, representative of traditional local architecture, was used as a pilot project to test the digitisation methodology.
The main steps involved in digitising this first door were as follows:

a) Stage 1: Data acquisition using photogrammetry

The first stage involved taking photographs of the monumental gateway, in accordance with good photogrammetry practice. Shots were taken from different angles and with sufficient overlap to enable the model to be reconstructed in 3D.

Figure 36: Door before treatment

b) Step 2: Processing the point cloud

The photographs were then processed using Geomagic Design software to generate a dense point cloud faithfully representing the geometry of the door. This step was used to clean up the cloud, eliminating extraneous elements and smoothing the surfaces.

Figure 37: Processing using Geomagic Design software

c) Stage 3: 3D modelling

The point cloud then imported into Geomagic Wrap software to generate a high-quality triangulated 3D mesh. This modelling made it possible to preserve the fine details of the architectural structure, while optimising the digital model for use future use.

Figure 38: Mesh generation using Geomagic Wrap software

d) Stage 4: Optimisation and export

Finally, once the 3D model had been cleaned up and simplified, it was exported in a standard format (.stl) for use in a variety of software and applications, including 3D printing and visualisation.

e) The technical challenges of digitisation

Although this first digitisation experiment validated the overall methodology, it also highlighted a number of technical challenges. The use a smartphone photogrammetry application proved insufficient to accurately capture the finest details of the door, particularly in the highest and most ornate parts. Despite post-processing efforts using Geomagic software, some areas with complex textures and traditional patterns could not be perfectly reconstructed in the final digital model. These difficulties highlighted the need to adapt data capture and 3D modelling techniques to meet the specific challenges posed by this type of richly detailed architectural heritage.

The lessons learned from this first experience have therefore guided the methodological choices for the next stage of the project to digitise the monumental gates of Meknes.

Untreated areas

Figure 39: The limits of smartphone photogrammetry

Conclusion

At the end of this second part, we have been able to demonstrate the immense potential of digitisation and 3D printing technologies in the field of cultural heritage preservation. The two case studies eloquently illustrate how these revolutionary innovations are making it possible to meet the major challenges facing heritage professionals.

Initially, implementation of reverse engineering using photogrammetry on the door of the Mohammed V museum in Rabat enabled us to generate a faithful 3D model, opening the way to numerous applications. By processing the point cloud using Geomagic software, we were able to produce a high-quality STL model, which was then optimised for 3D printing using the SLA method. This approach was enhanced by the laser scan of the Colosseum in Rome, providing an enlightening comparison of the two 3D capture techniques. As well as simply providing a physical reconstruction, these digital models are invaluable tools for studying, restoring and disseminating our heritage.

Secondly, the creation of a digital library of the monumental gates of Meknes is an ambitious initiative aimed at preserving this rich architectural heritage. Photogrammetric scanning of the gateway to the Riad Golden Mansour, followed by processing using Geomagic Design and Wrap software, has produced a high-quality STL model that is ready for a wide range of uses. This approach paves the way for the long-term preservation of these heritage treasures, while facilitating their accessibility and in-depth study.

Through these two projects, we were able to highlight the incredible added value that digitisation and 3D printing technologies bring to the preservation of cultural heritage. These innovations offer new perspectives in terms of conservation, restoration and dissemination, helping to ensure that these ancestral legacies are passed on to future generations. Far from being confined to simple restoration tools, these technologies are proving to be real levers for enhancing the value and understanding of our heritage.

Part 2

Building a robotic arm

Robotics is a rapidly expanding field, offering many innovative applications in a variety of sectors. As part of this project, we took on the challenge of designing and building a prototype 6-axis robotic arm, a powerful tool capable of performing a wide range of tasks with precision and dexterity. This first part of the report details the key stages in the design, programming and implementation of this innovative robotic system, highlighting the technical choices made, the challenges encountered and the performance achieved.

Introduction

In industrial automation, manipulator robots with articulated arms play an essential role in carrying out complex tasks such as handling, assembly and machining. These robots offer great flexibility and precision of movement, making it possible to increase the productivity and quality of manufacturing processes. The aim of this project is to design and build a prototype 6-axis industrial robot capable of moving a given load. Unlike conventional industrial robot arms, which are often manufactured using traditional production techniques, our approach involves using additive manufacturing technologies, more specifically 3D printing, to produce the robot's mechanical parts. This innovative approach offers a number of advantages, particularly in terms of reducing manufacturing costs and lead times, while taking advantage of the possibilities offered by additive manufacturing processes.
The main technical challenges of this project are as follows:

- Adapt the initial design of the robot for manufacture using 3D printing;
- Select the most suitable materials for the robot parts in terms of mechanical, thermal and wear resistance constraints;
- Define the optimum 3D printing parameters to guarantee the robustness and reliability of the parts;
- Integrate the robot's electrical and electronic control and motorisation systems;
- Validate the robot's performance in terms of load capacity, positioning accuracy and repeatability;

This report begins by presenting the different types of industrial robotic arm already on the market. Secondly, it details the stages involved in producing our prototype robotic arm, by discovering the additive manufacturing process and using the machines and materials available to us, as well as the future prospects for the project. [7]

I. Existing industrial robotic arms

Before presenting the development of our robotic arm prototype, it's important to take stock of the various industrial robotic solutions already on the market.
Industrial robotic arms fall into several categories, depending mainly on their mechanical architecture and field of application. These include
Cartesian robots
Cartesian robots, also known as gantry robots, are characterised by an architecture consisting of three orthogonal linear axes. They offer high positioning accuracy and are particularly well suited to tasks requiring high repeatability such as machining, assembly or palletising. Their working area is generally limited to a parallelepiped space.

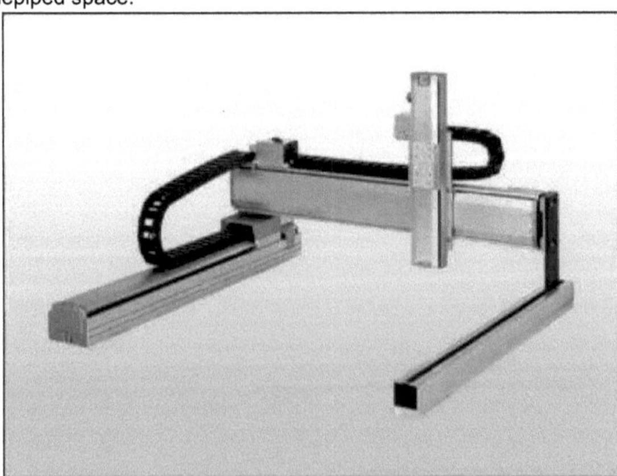

Figure 40: Cartesian robots

- **Anthropomorphic robots**

Anthropomorphic robots, inspired by the morphology of the human arm, have a 6-axis rotation architecture for great dexterity and flexibility of movement. They are widely used in assembly, welding and painting applications, where their ability to access hard-to-reach areas is an asset.

- **SCARA robots**

SCARA (Selective Compliance Assembly Robot Arm) robots feature a structure composed of two horizontal axes of rotation and one vertical axis of translation. They offer high rigidity and are suitable for insertion, pick-and-place or micro-assembly tasks requiring high precision in a horizontal plane.

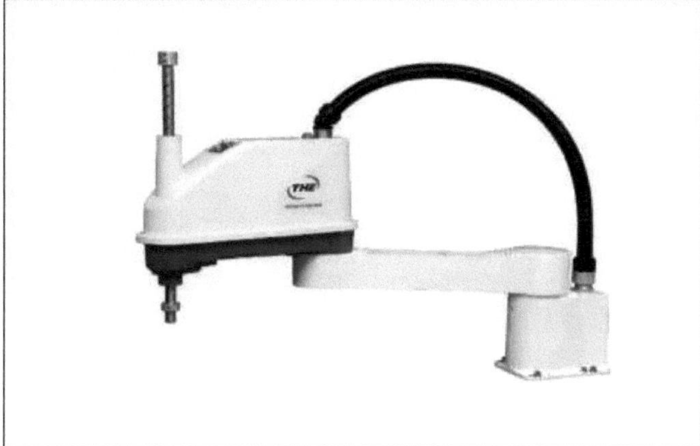

Figure 41:: SCARA robot

- **Delta robots**

Delta robots, with their parallel structure, stand out for their high speed and excellent repeatability. They are widely used in high-speed packaging and pick-and-place applications.

Each of these robotic architectures has its advantages and disadvantages in terms of dexterity, payload, working area and speed of movement. The choice of the most suitable type of robot therefore depends very much on the specifics of the industrial application in question.

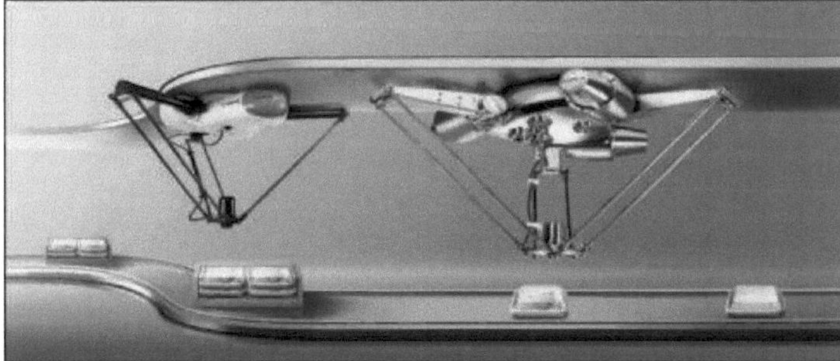

Figure 42: Delta robot

II. Initial design of the robot

Our prototype robotic arm is fundamentally different from conventional industrial models. While the latter are generally manufactured using traditional production methods, we have chosen exploit additive manufacturing technologies, more specifically 3D printing, to produce the mechanical components of our system.

This choice was motivated by time constraints, which did not allow us to carry out a complete design of the robot from the outset. We therefore decided to base our design on an existing architecture, namely a 6-axis mechanical structure. This 6-degree-of-freedom configuration offers great flexibility and dexterity of movement, essential characteristics for covering a wide range industrial applications. In the following sections, we will detail the composition of this robotic arm, focusing on the choice of materials and 3D printing parameters selected to guarantee the robustness and reliability of our prototype.

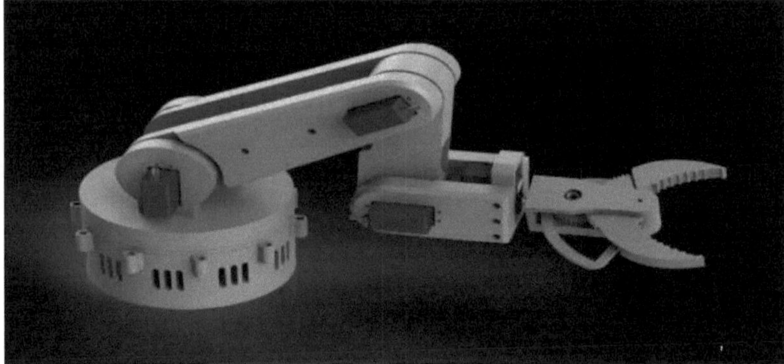

Figure 43: Robot design

1. Mechanical design of the robotic arm

Our robotic arm consists of :

> **End effector: Gripper**

Our robotic arm is equipped a gripper, end effector of the system. This gripper, programmed via an Arduino interface, can be used to grasp and move a variety of objects with precision thanks to its two motorised jaws. The design of the gripper has been optimised to provide a clamping force suitable for handling a wide range objects of different sizes and shapes, making the robot versatile for many handling tasks.

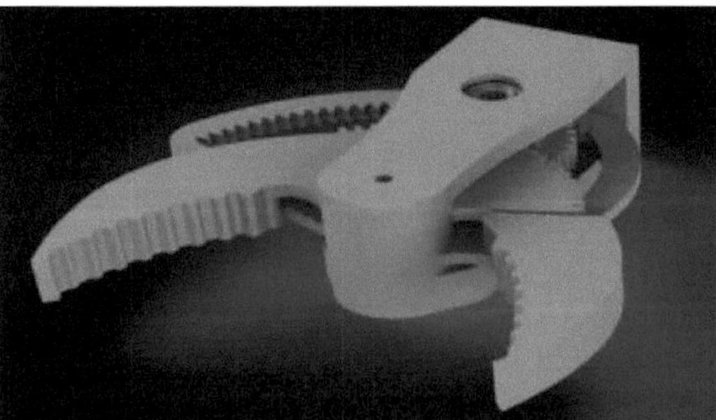

Figure 44: Robot gripper

> **Articulated segment**

The articulated part of the robotic arm is fitted with 5 high-performance servo motors, giving the whole unit great mobility. This architecture ensures greater dexterity of movement, enabling the robot to position and orientate the end-effector with great flexibility in the workspace. The joints are built using 3D printed parts to ensure that the assembly is both lightweight and robust.

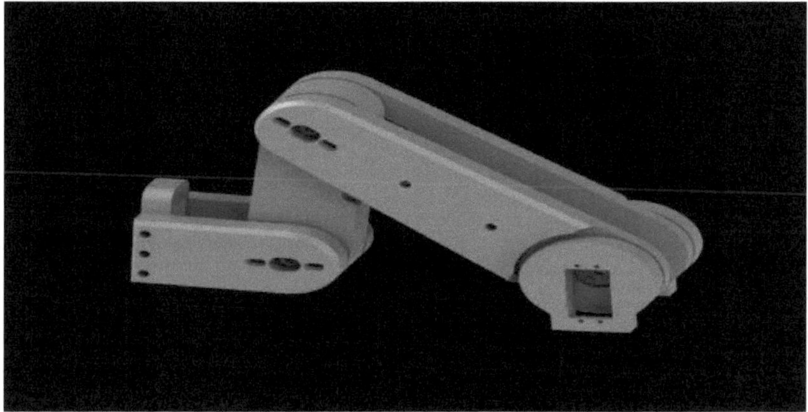

Figure 45: Articulated segment

> **Rotating base**

The robotic arm rests on a rotating base, allowing full 360-degree rotation. This motorised base is also manufactured using 3D printing and incorporates a gear system to transmit the rotational movement. This feature significantly increases the workspace accessible to the robot, optimising its range of action and versatility of use for a wide variety of industrial tasks.

Figure 46: Rotating base

2. Electronic system components

> **Arduino Mega board: the central controller**

The heart of the electronic system is an Arduino Mega board, which acts as the central controller for all the components of the robotic arm. It offers a wide range of inputs/outputs and high computing power.

This Arduino board coordinates the various actuators and sensors, enabling fine, precise management of the arm's movement.

> **Servomotors: joint actuators**

Six high-performance servomotors are integrated into the robotic arm, each responsible for the movement of a joint.

These servomotors allow precise control of the angular position of each segment, contributing to the dexterity and flexibility of the entire system.

> **Stepper motors: rotary base actuators**

In addition to the servomotors, a stepper motor is used to rotate the base of the robotic arm. This type of actuator offers high positioning accuracy, which is essential to guarantee smooth, controlled rotation of the platform.

> **Servo pilot: actuator control (PCA card)**

To control the servomotors and stepper motor, a dedicated servo-pilot is integrated into the system. This electronic module converts the control signals from the Arduino board into motion commands that can be used by the various actuators.

> **Battery: power supply**

A rechargeable battery powers the entire electronic system, ensuring the robotic arm's autonomy and mobility. This energy source eliminates the need for a wired connection, giving greater freedom of movement and experimentation.

> **Wiring and breadboard: interconnections**

A wiring network linked to a breadboard facilitates connections between the various electronic components. This modular approach simplifies the assembly and maintenance of the system, while offering great flexibility for future modifications.

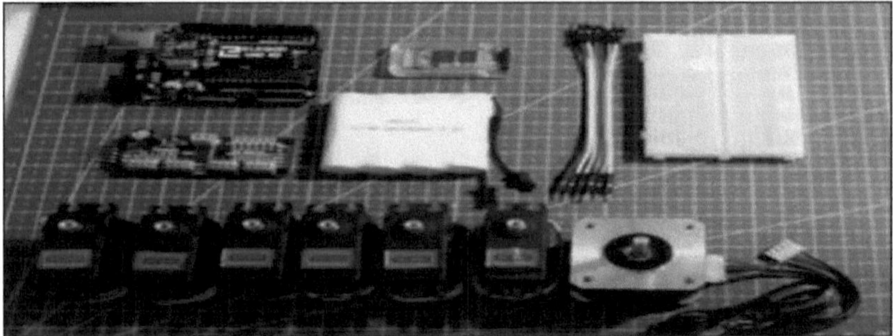

Figure 47:Electronic components of the system

111. Robot made using additive manufacturing

1. The additive manufacturing process

Additive manufacturing, also known as 3D printing, is a set of processes used to create physical parts from digital models. This manufacturing approach differs from traditional methods by adding material progressively, layer by layer, rather than removing material as in machining.
The main additive manufacturing processes are as follows:

- **Tank photopolymerisation:** This process uses a bath photosensitive resin which is selectively hardened by a light beam (laser or screen) according to the 3D model. The part is built up layer by layer by progressively immersing the manufacturing tray in the tank.
- **Material spraying**: This process consists of selectively spraying fine droplets of material (generally a polymer) onto a plate, which are then hardened by heat treatment or radiation.
- **Concentrated energy material deposition:** In this method, a laser beam or electric arc locally melts a metal wire or polymer, depositing the material layer by layer to form the part.
- **Layer lamination:** This process involves cutting the 3D model into thin slices, which are then stacked and glued together to progressively build up the final part.
- **Binder spraying:** A print head selectively deposits a binder on a bed of powder (polymer, metal or ceramic), which then agglomerates the desired areas.

- **Powder bed fusion:** A laser or electron beam melts a thin layer of powder (polymer, metal or ceramic) locally to create the geometry of the part.
- **Molten wire deposition:** This process involves extruding a molten thermoplastic wire layer by layer to progressively build up the part.

To produce the parts for our robotic arm, we chose use one of the additive manufacturing processes described above. More specifically, we chose fused deposition technology, or FDM. [8] This 3D printing method proved to be particularly suitable for manufacturing plastic components such as the arm's chassis, supports and attachments.

To do this, we used 3D printers equipped with an extrusion head that heats the filament material until it melts. The extrusion head then moves according to the coordinates defined by the digital model, depositing the molten material layer by layer to gradually build up the object.

As far as the raw materials used are concerned, we have selected three types of thermoplastic material: PETG, PLA and PLA+.

The choice of material was based on the specific properties and constraints of each component:

- **PETG** (Polyethylene glycol terephthalate): For parts requiring high mechanical strength and rigidity. This material excellent mechanical properties, good toughness and impact resistance.
- **PLA** (Polylactic Acid): For fasteners and less critical parts, we have favoured the use of PLA. This bioplastic has the advantage of being biodegradable, while offering good performance in terms of print resolution and surface finish.
- **PLA+** : This improved grade of PLA has been selected for parts requiring greater mechanical strength than standard PLA, while retaining the benefits of this environmentally-friendly material. PLA+ was used for certain intermediate parts of the robotic arm.

This judicious choice of materials according to the specifications of each component has enabled us optimise the performance and reliability of the entire robotic system. [9], [10]

2. The production stages

- Having defined the process used, the material and the machinery, here are the stages in the manufacture of our prototype [11] :

a. Preparation of CAD files in STL format

- We already had the CAD files for the various components. We exported them in STL format. This is a crucial step, as it transforms our digital models into files that are compatible with the 3D printing process.
- The STL format generates the meshing information needed to define the geometry of each part, preparing us the next stage of print preparation.

Figure 48: CAD file for the rotating base

b. Preparing printing parameters

- To prepare the printing parameters, we opted to use Ultimaker Cura software. The orientation of each part on the printing plate was carefully planned, taking into account various criteria. We sought to minimise printing time while reducing the use of substrates, thereby reducing post-processing time and preserving surface quality. As far as supports are concerned, we placed them on all cantilevered areas with an angle to the vertical plane of less than a certain threshold, generally recommended at 45° for our printers.

In addition, we have chosen an appropriate layer thickness depending on the desired level of finish for the surfaces.

- Once all these parameters have been defined, the final step is cut the STL model into slices using the "Slice" tool, thus preparing the path instructions for 3D printing.

Figure 49: Importing the CAD model to the slicer and adjusting the print parameters

c. Start printing

- We select the essential parameters to ensure quality printing. Firstly, we carefully choose the temperature of the printing plate, which plays a crucial role in the adhesion of the material and the stability of the part during the printing process. In addition, we determine the appropriate print speed, taking into account the complexity of the part, its size and the surface quality required. By adjusting the print speed, we optimise the balance between production speed and accuracy of detail.

By combining these parameters thoughtfully, we ensure that every print is produced accurately and reliably, meeting the highest quality standards for our prototype.

Figure 50: Piece after printing

4. Post-treatment

In the final stage of the manufacturing process, we post-process our prototype. Once printing is complete, we carefully remove the supports used to stabilise the cantilevered areas during printing. This step is crucial to achieving a clean, smooth final finish on our parts. By methodically removing the supports, we ensure that we do not damage the details or surfaces of the part. We use appropriate tools, such as pliers and precision knives, to remove the backing efficiently while preserving the integrity of our parts. Once this post-treatment stage is complete, our parts are ready to be assembled to form the final prototype, representing the fruit of our labour and meticulous engineering.

3. Assembling the 3D-printed parts

Once the various components of the robotic arm have been manufactured using 3D printing, the next step is to assemble them to form the complete system.

This assembly phase involved the following elements:

Surface preparation :

The contact surfaces between the parts have been carefully cleaned and prepared to ensure optimum adhesion during assembly.

Assembly methods :

Depending on the design of each joint, we have used different assembly techniques, such :

Screwing: Tapped holes were used to secure the parts using screws.

Interlocking: Some parts have been fitted directly into each other by means of flush-mounting systems.

Bonding: For permanent assemblies, we structural adhesives adapted to plastic materials.

We also incorporated moving parts to facilitate certain movements of the robotic arm. For example, we added ball bearings to the arm's key joints. These bearings ensure smooth rotation and a significant reduction in friction, improving the system's manoeuvrability and performance.

In addition, for the base of the robotic arm, we designed and 3D printed balls that fit into appropriate housings. This system of printed 'bushes' makes it easier to rotate the base, giving the whole arm greater mobility. The successful integration of these mobile elements, in addition to conventional assembly techniques, has helped to optimise the overall mechanical design of the robotic arm and improve its functionality.

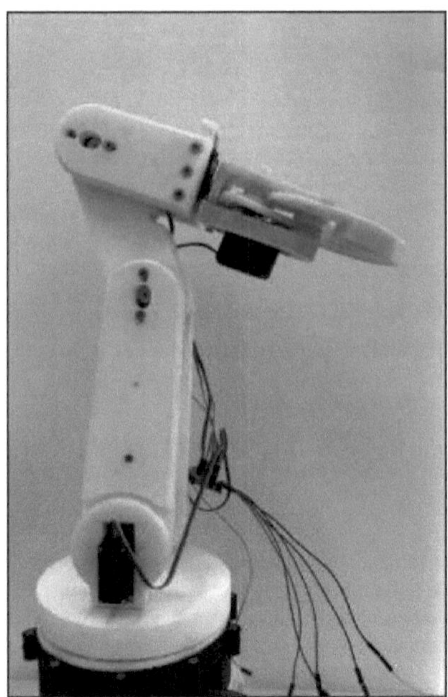

Figure 51: Final result of the robot

4. Arduino code and wiring

To bring our prototype 6-axis robotic arm to life, we developed a control programme based on the Arduino platform. This electronic solution enabled us to precisely control the robot's various movements and ensure smooth coordination.

The choice of the Arduino platform proved to be a wise one, given its flexibility, ease of programming and accessibility. Thanks to its many digital and analogue inputs/outputs, we were able to interface easily with the robot's various electronic components, such as the servo motors and stepper motor.

The design of the Arduino code was a key element in the development of our prototype.

A robust software architecture had to be put in place, capable of managing the coordinated movements of the 6 axes of the robotic arm in real time. This required in-depth consideration of inverse kinematics, feedback loops and performance optimisation.

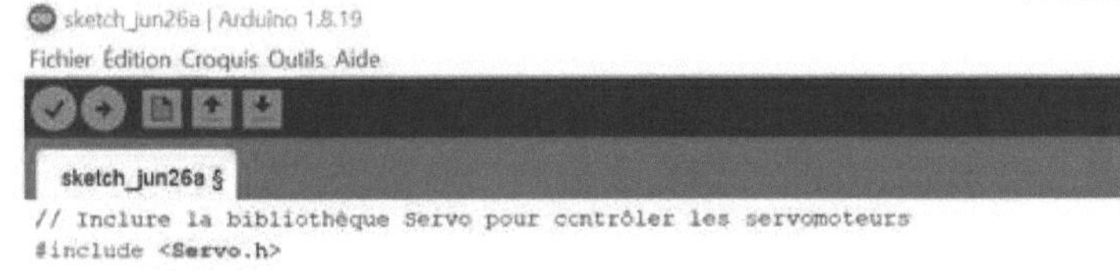


```
// Inclure la bibliothèque Servo pour contrôler les servomoteurs
#include <Servo.h>

// Déclarer les 6 servomoteurs
Servo servo1, servo2, servo3, servo4, servo5, servo6;

void setup() {
  // Attacher les servomoteurs aux broches digitales correspondantes
  servo1.attach(3);
  servo2.attach(5);
  servo3.attach(6);
  servo4.attach(9);
  servo5.attach(10);
  servo6.attach(11);
}

void loop() {
  // Faire tourner les servomoteurs 1 et 2 de 30 degrés
  servo1.write(30);
  servo2.write(30);
  delay(1000); // Attendre 1 seconde

  // Faire tourner le servomoteur 3 de 60 degrés
  servo3.write(60);
  delay(1000);

  // Faire tourner le servomoteur 4 de 100 degrés
  servo4.write(100);
  delay(1000);

  // Faire tourner le servomoteur 5 de 360 degrés
  servo5.write(360);
  delay(1000);

  // Faire tourner le servomoteur 6 de 90 degrés, puis le ramener à 0 degré
  servo6.write(90);
  delay(1000);
  servo6.write(0);
  delay(1000);

  // Répéter la séquence
}
```

Conclusion

At the end of this project to develop a robotic arm, we went through the various key stages to design and produce a working prototype.

Firstly, we designed the mechanical structure of the arm using additive manufacturing (3D printing). This enabled us to create a light, robust framework capable of supporting the movements of the various joints. The judicious choice of servomotors and their careful integration into the structure ensured that the movements were fluid and precise.

We then developed the control system using an Arduino board. The Arduino code we wrote coordinates the activation of the servomotors to faithfully reproduce the movements desired by the user. The electronic wiring has been carefully designed to guarantee the reliability of the whole system.

Although this prototype is fully functional, we plan add a new feature to make it even easier to use. Using additive manufacturing, we plan to design and manufacture control gloves that will enable the user to operate the robotic arm intuitively, by naturally reproducing his hand movements.

This new feature will greatly improve the ergonomics and user experience of the robotic arm, making it even more accessible and versatile. Although this prototype is already very advanced, we will continue to work on its constant improvement to make it an even more effective robotic tool, adapted to the needs of users.

General conclusion

At the end of this ambitious project report, we can proudly say that the objectives set have been fully achieved, offering remarkable advances in two crucial areas: robotics and cultural heritage.

The first part of the report explored the tremendous potential of digitisation and 3D printing technologies for safeguarding cultural heritage. Through two remarkable case studies, we were able to demonstrate how these revolutionary innovations are making it possible to meet the major challenges facing those involved in heritage conservation. Reverse engineering using photogrammetry, followed by optimisation and 3D printing, has made it possible to generate faithful and versatile digital models of heritage items. Similarly, the creation of a digital library of the monumental gates of Meknes is an ambitious initiative aimed at preserving this rich architectural heritage. These initiatives offer new perspectives in terms of conservation, restoration and dissemination, contributing to greater accessibility and a deeper understanding of our treasures from the past.

Through these two complementary projects, this project report highlights the synergy between the disciplines of robotics and digital technologies applied to cultural heritage. These innovations represent real levers for meeting the challenges of preserving, enhancing and passing on this priceless heritage to future generations. The results obtained demonstrate the power of the interdisciplinary approach, paving the way for exciting new prospects in this crucially important field.

The second chapter was devoted to the design and production of a six-axis robotic arm capable of lifting a load of 500 grams. This technical feat required the application of cutting-edge skills in mechanics, electronics and programming. The judicious choice of components, precision engineering and sophisticated control algorithms have resulted in a reliable, easy-to-handle, high-performance robotic system. This achievement represents a significant contribution to new approaches to heritage preservation and handling, paving the way for future applications in delicate or inaccessible environments.

> The machines used in FDM :

Figure 53: Anet printer Figure 55: Creality A10 max

Figure 52: Artillery printer Figure 54: Artillery printer

The machines used in SLA :

Figure 56 :Elegoo Jupiter SE

The different parts that make up the robot :

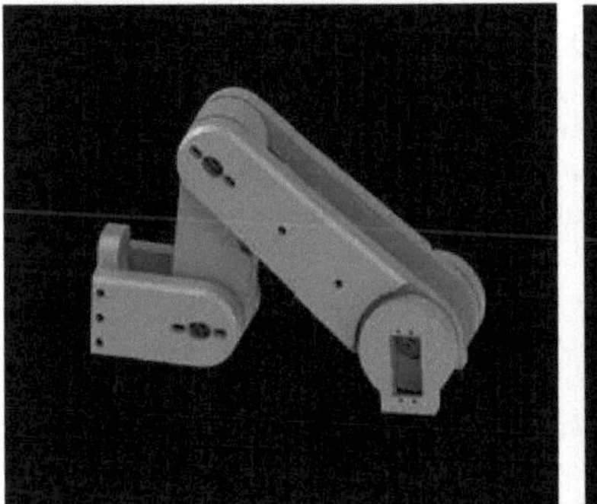
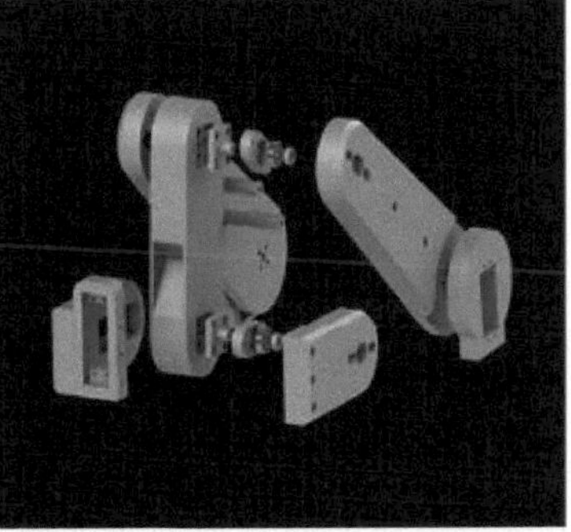

Figure 58: The robot arm

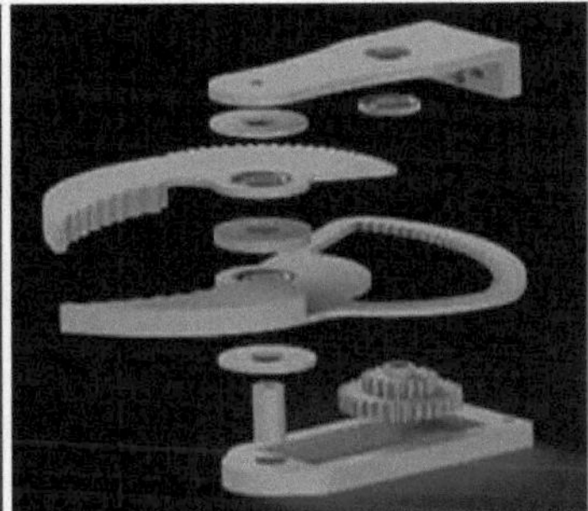

Figure 59: Gripper

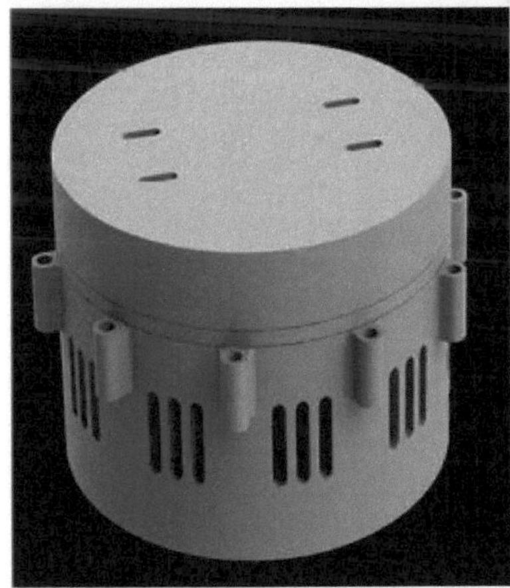

Figure 60: Rotating base

[1]- BENASSA EL FAHIME, *:* " *Numerisation 3D Pretretement Des Nuages De Points* ", - *Diplome des études superieures approfondies*, DESA 2007.
[2]- RADOUANI M., SAKA A., BELKEBIR H., EL FAHIME B., CARRARD M., *""Pre-processing of point clouds: Comparative study of commercially available software".* 4th International Conference on Integrated Design and Production, CPI'2005, Meknès - Morocco, CDRom paper, 11 pages, October 2005.
[3]- LINDsAY MACDONALD, "*Applying Digital Imaging to Cultural Heritage*". Handbook, Université Parissaclay, 83 pages, 2006.
[4]- R. ROUssEL ET L. DE LUCA" *Une Approche Pour Construire Un Rapport Numerique Complet De La Cathedrale Notre Dame Apres L'incendie, En Utilisant La Plateforme AIOLI* ". article consulted at: https:*//isprs-archives.copernicus.org/articles.*
[5]- ISO 52910:2018, "*Additive manufacturing - Design - Requirements, guidelines and recommendations*".
[6] - INssAF BAHNINI, *"Méthodologie de prise en compte des incertitudes et des interactions des caractéristiques produit/proccess. Application aux procédés de fabrication additive*". Thesis, Université ABDELMALEK EssAADI, 2021.
[7]- T. VARADY, R. MARTIN, J. COX "*Reverse Engineering of geometric models - an introduction, Computer-Aided Design*". vol 29(4), pp. 255-268.
[8]- MATTHIAs BORDRON, *"Modelling and calibration for digitisation robotised*". Thesis, Université Parissaclay, 2019.
[9]- ISO 52900:2015, *"Additive manufacturing - General principles Terminology*".
[10]- ISO 17296-2:2015, "*Additive manufacturing - General principles. Part 2: Overview of process categories and raw materials*".
[11]- INsAF BAHNINI, MICKAEL RIVETTE, AHMED RECHIA, ALI sIADAT, ABDELILAH ELMEsBAHI, "*Additive manufacturing technology: the status, applications, and prospects*". The International Journal of Advanced Manufacturing Technology (201 8) 97:147-161

Printed by Books on Demand GmbH, Norderstedt / Germany